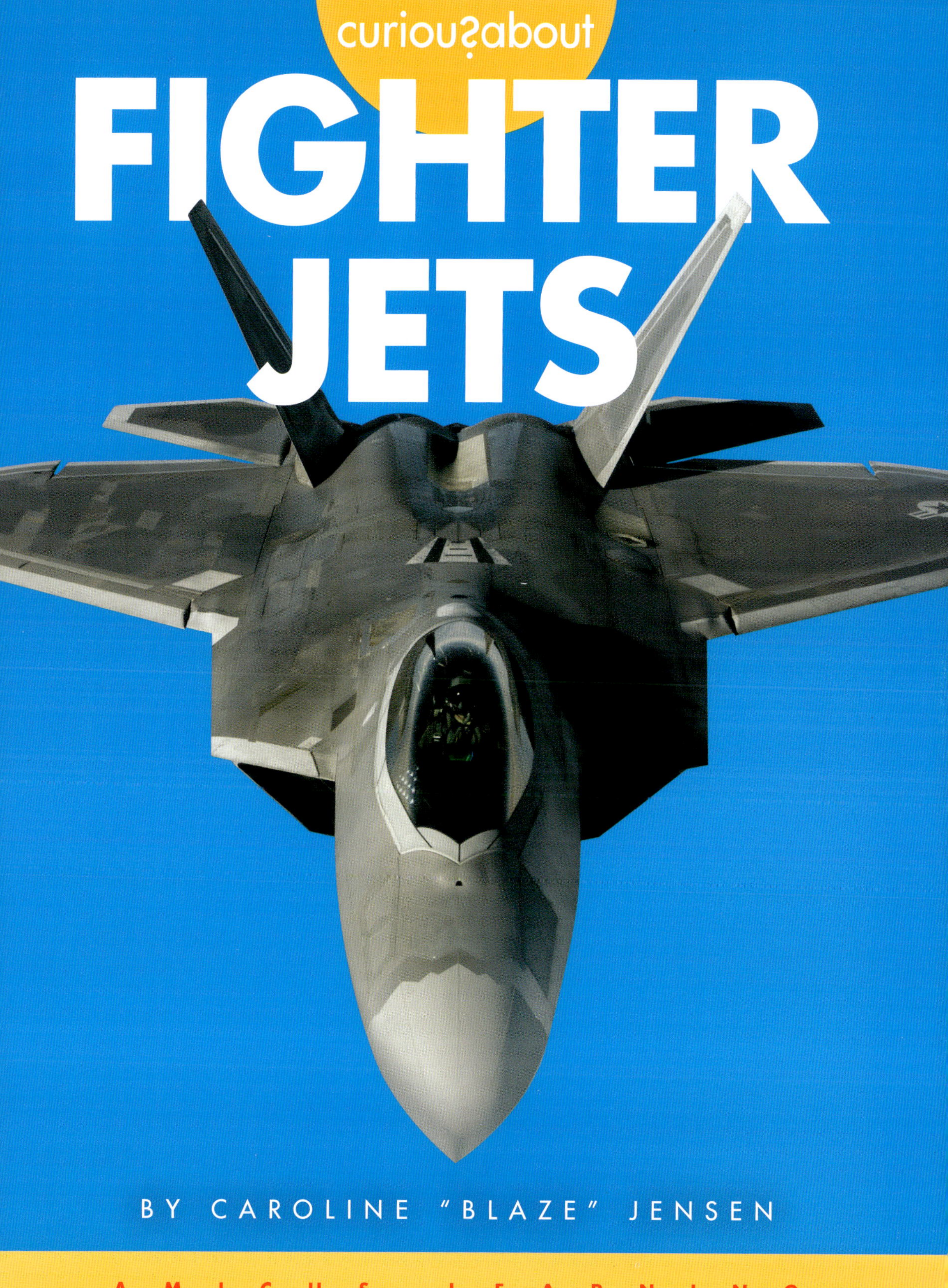
curious?about
FIGHTER JETS
BY CAROLINE "BLAZE" JENSEN
AMICUS LEARNING

What are you

curious about?

Curious About is published by
Amicus Learning, an imprint of Amicus
P.O. Box 227
Mankato, MN 56002
www.amicuspublishing.us

Editor: Alissa Thielges
Series Designer: Kathleen Petelinsek
Book Designer: Aubrey Harper

Library of Congress Cataloging-in-Publication Data
Names: Jensen, Caroline, author.
Title: Curious about fighter jets / by Caroline Jensen.
Description: Mankato, MN : Amicus Learning, [2025] | Series: Curious about military machines | Includes bibliographical references and index. | Audience: Ages 5–9 | Audience: Grades 2–3 | Summary: "Early elementary readers learn how fighter jets work in this inquiry based nonfiction book about their size, speed, and battle strength. Includes infographics and back matter to support research skills, plus table of contents, glossary, and index"—Provided by publisher.
Identifiers: LCCN 2023038612 (print) | LCCN 2023038613 (ebook) | ISBN 9781645493167 (library binding) | ISBN 9781645494041 (ebook)
Subjects: LCSH: Fighter planes—United States—Juvenile literature.
Classification: LCC UG1242.F5 J467 2025 (print) | LCC UG1242.F5 (ebook) | DDC 623.74/6440973—dc23/eng/20231103
LC record available at https://lccn.loc.gov/2023038612
LC ebook record available at https://lccn.loc.gov/2023038613

Photo credits: Airforce 17; DVIDS/94th Airlift Wing 17, Chief Petty Officer Shannon Renfroe 12–13, 18–19, Cpl. Joseph Abrego 17, Kyra Helwick 14–15, Lance Cpl. Samantha Rodriguez 17, Staff Sgt. Alan Ricker 20–21, Master Sgt. John Nimmo, Sr. 9, Master Sgt. Nicholas Priest 4–5, 14–15, Ralph Branson 9, Senior Airman Andrew Sarver 11, Senior Airman Devante Williams 8, Senior Airman Duncan Bevan 7, 9, Staff Sgt. Codie Trimble 10, Staff Sgt. Justin Parsons 9, Staff Sgt. Michael Battles 6, Staff Sgt. Samantha Mathison 17, Tech. Sgt. Jason Robertson cover, 9; Noun Project/shashank singh 22–23 (icons)

Printed in China

CHAPTER ONE

What is a fighter jet?

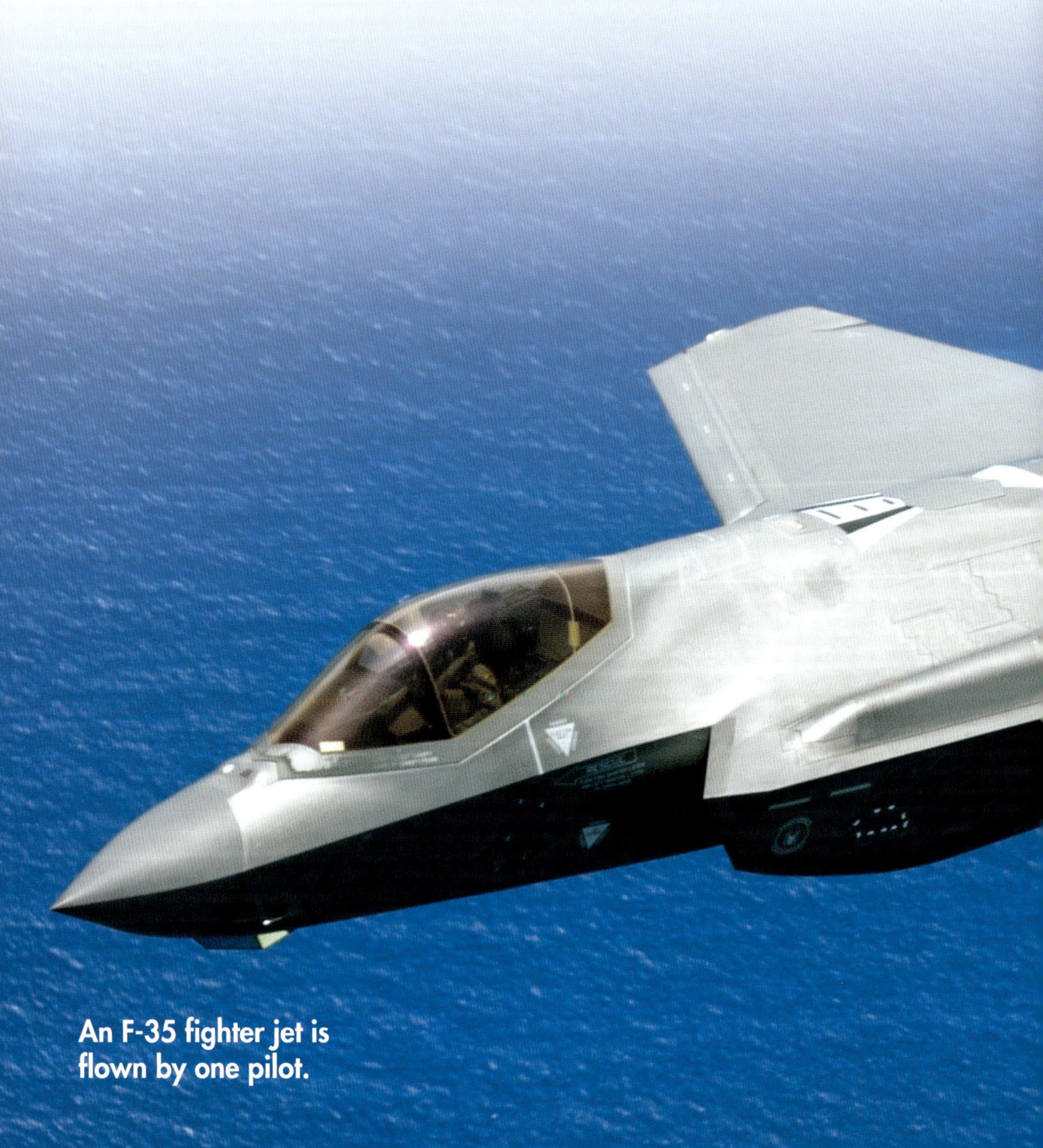

An F-35 fighter jet is flown by one pilot.

A powerful aircraft. Fighter jets defend our country. They fly **supersonic** speeds and make tight turns. Some only have one pilot. These jets are loaded with weapons. They rule the skies with speed and **stealth**. Technology hides them from enemies.

DID YOU KNOW?
The first American fighter with a jet engine was the P-80 Shooting Star in 1945.

What weapons do fighter jets have?

So many! Fighter jets take down enemy planes with missiles. They shoot at the enemy with guns. They can also drop bombs. Fighter jets have special abilities to see at night. They can also carry extra **fuel** in tanks.

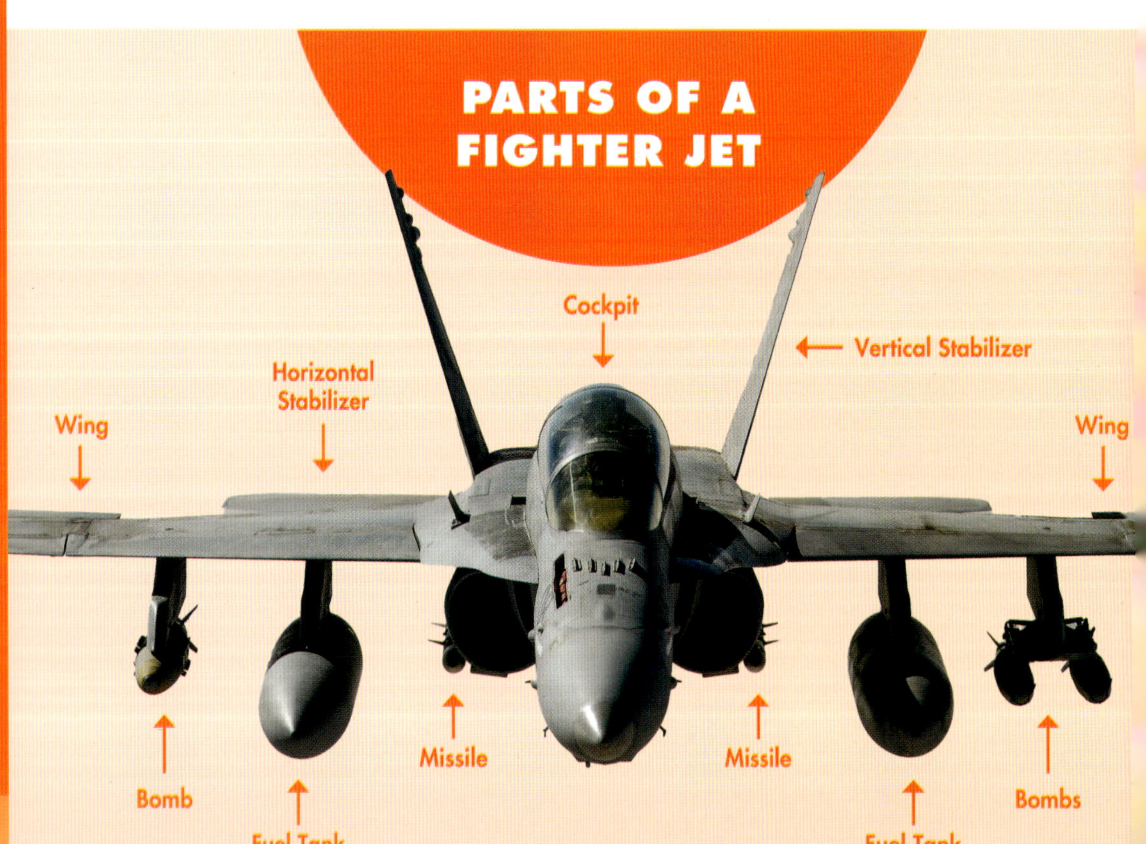

Weapons can be hidden on some jets when they are not in use.

What branches of the military use fighter jets?

F-35s can take off and land vertically.

Three U.S. military branches use them. The Air Force flies the most. It has over 1,300 fighter jets! The Navy has about 1,000 fighter jets. The Marines use a special kind of F-35. Both the Navy and Marines land jets on huge aircraft carriers.

F-15
EAGLE

F-16
FIGHTING
FALCON

F-18
HORNET

F-22
RAPTOR

F-35
LIGHTNING II

CHAPTER TWO

How fast does a fighter jet go?

Fighter jets sometimes create a vapor cloud at high speeds.

Faster than sound! Fighter jets break the **sound barrier**. Going faster than **Mach** 1 makes a loud boom. Fighters use Mach to measure how fast they can go. Most fighters today can go Mach 2 or more. That's twice as fast as sound travels through the air!

A jet pilot can feel nine times their weight in pressure.

DID YOU KNOW?

The fastest jet was the North American X-15 in 1967. It went Mach 6.7, or about 75 miles (121 kilometers) a *minute*!

Afterburner makes flames shoot out the back of a fighter jet.

How does a fighter jet fly so fast?

DID YOU KNOW?

Fighter jets have **radar** in the nose of the aircraft. The radars guide missiles to enemy planes.

It has a powerful engine. It burns fuel to heat and expand air. The hot air is pushed out the back to create **thrust**. The jets also have an afterburner. This device adds extra fuel to the hot air for a boost of speed.The jet climbs, dives, and rolls super fast.

How long can a fighter jet stay in the air?

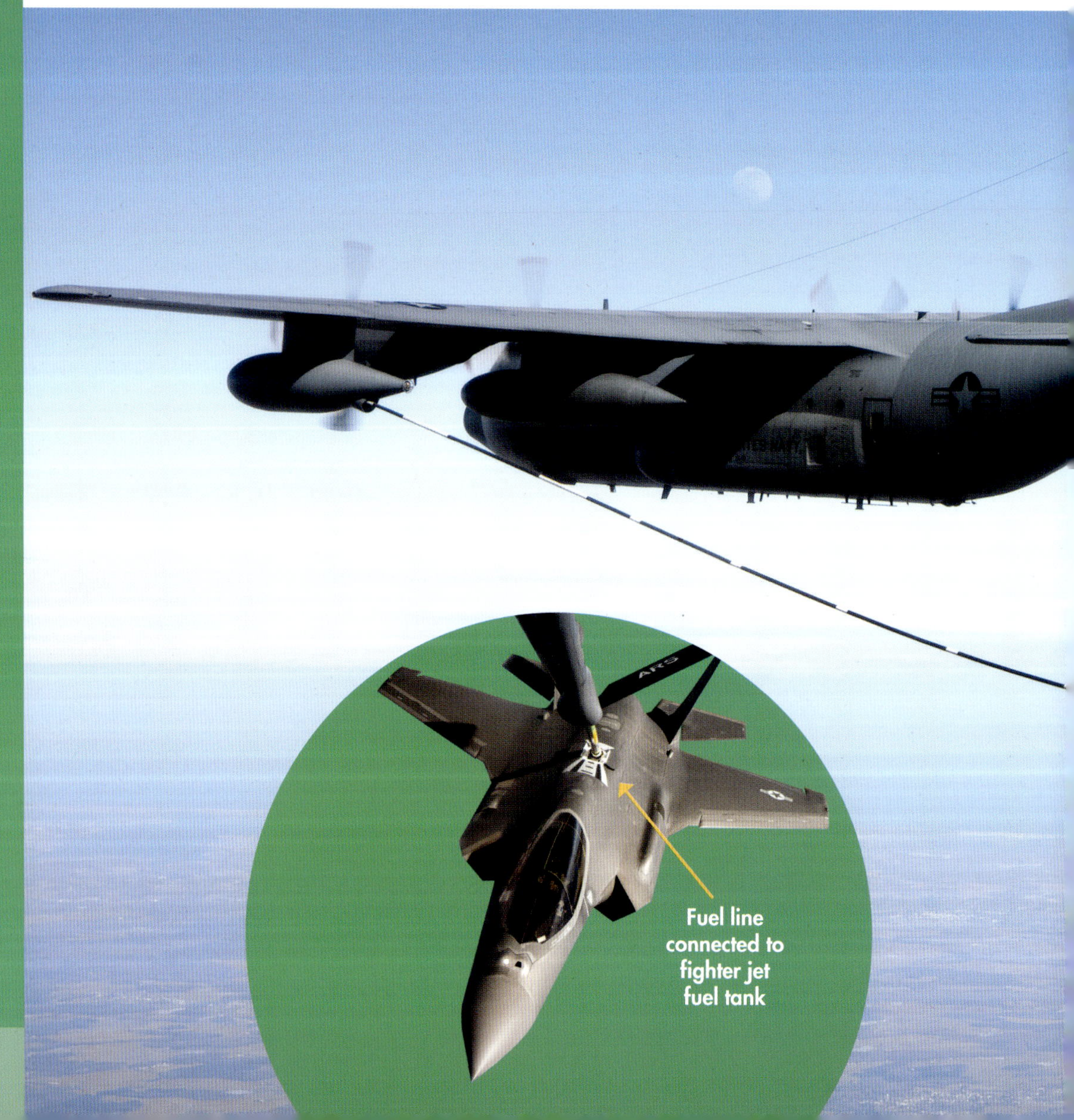

Fuel line connected to fighter jet fuel tank

It depends. The fuel goes fast when they fly low. Those flights are shorter than an hour. If a jet flies higher, it can cruise at a slower speed. Then it can go more than two hours on a full tank. But jets don't need to land to refuel. They can hook up to a tanker airplane in the air.

Some tanker planes can refuel two fighter jets at a time.

What is it like to fly a fighter jet?

Loud and hot! The jets are very noisy and shake a lot. Fighter pilots need steady hands and **endurance** for long flights. Pilots bring water and special plastic bags with them. If they need to pee, they go in a special plastic bag.

Pilots train hard to handle the challenges of flying a jet.

FIGHTER PILOT GEAR

G Suit and Life Preserver

Helmet and Mask

Harness

Does flying a fighter jet look like the movies?

A Navy pilot runs practice drills in an F-35 in California.

Not really. In movies, fighter jets fly close to each other. In real life, pilots spot other planes from far away with their eyes or radar. Fighter jets fly in groups called formations. But they are spread out so it's harder for enemies to see them coming.

DID YOU KNOW?
An "Ace" is a fighter pilot that shoots down five or more enemy airplanes.

How can I see fighter jets up close?

Thunderbirds fly F-16 Fighting Falcons during shows.

Head to an air show! You can see the U.S. Air Force Thunderbirds or the U.S. Navy Blue Angels. Both teams travel around the United States and other countries. They perform amazing **maneuvers**. They show all the extreme ways a fighter jet can be flown.

ASK MORE QUESTIONS

How much does a fighter jet cost?

What does it take to become a fighter pilot?

Try a BIG QUESTION: How have fighter jets changed modern warfare?

SEARCH FOR ANSWERS

Search the library catalog or the Internet.
A librarian, teacher, or parent can help you.

Using Keywords
Find the looking glass.

Keywords are the most important words in your question.

If you want to know:

- the cost of a fighter jet, type: FIGHTER JET COST
- how to become a pilot, type: FIGHTER JET PILOT TRAINING

FIND GOOD SOURCES

Here are some good, safe sources you can use in your research.
Your librarian can help you find more.

Books

Fighter Jets
by Wendy Hinote Lanier, 2019.

Fighter Jets in Action
by Mari Bolte, 2024.

Internet Sites

National Museum of the Air Force
www.nationalmuseum.af.mil
This official website for the Air Force museum has information on every military plane.

Smithsonian Air and Space Museum
airandspace.si.edu
The Smithsonian is the biggest U.S. national museum. This site has information on airplanes of all kinds as well as spacecraft.

Every effort has been made to ensure that these websites are appropriate for children. However, because of the nature of the Internet, it is impossible to guarantee that these sites will remain active indefinitely or that their contents will not be altered.

SHARE AND TAKE ACTION

Ask an adult to take you to an airport near your home and ask questions.
People in aviation love to share their passion with kids!

Learn to fly!
Sign up for a learn-to-fly flight.

Flying is expensive but there are scholarships available.
Look up scholarships and camps for kids in aviation.

GLOSSARY

endurance The ability to do something difficult for a long time.

fuel A material, such as gas, that is burned to produce heat or power.

Mach A measure of high speed compared to the speed of sound, which is Mach 1.

maneuver A clever or skillful action or movement.

radar A device that sends out radio waves to find the position and speed of a moving object.

sound barrier A large increase in air resistance that occurs as an aircraft nears the speed of sound.

supersonic Faster than the speed of sound.

stealth A secret and quiet way of moving or behaving.

thrust To push forward with force.

INDEX

About the Author

Caroline "Blaze" Jensen lives in Wisconsin with her son, Finn, and dog, Gunner. She flew 3,600 hours in the Air Force as a fighter pilot including F-16 combat missions and Thunderbirds demonstrations. She loves sharing her passion for flying with kids of all ages.